I0797692

FUNCTIONS & VARIABLES

by Teddy Borth

Cody Koala

An Imprint of Pop!
popbooksonline.com

0011100100010011100100100110010100

abdobooks.com
Published by Pop!, a division of ABDO, PO Box 398166, Minneapolis, Minnesota 55439.

Printed in the United States of America, North Mankato, Minnesota

052021
092021

THIS BOOK CONTAINS RECYCLED MATERIALS

Cover Photo: Shutterstock Images
Interior Photos: iStockphotos, 5, 9 (top), 9 (bottom left), 11; Shutterstock Images, 6, 9 (bottom right), 13 (top), 13 (bottom left, center, and right), 15, 15, 17, 18, 21

Editor: Elizabeth Andrews
Series Designer: Laura Graphenteen

Library of Congress Control Number: 2020948272

Publisher's Cataloging-in-Publication Data
Names: Borth, Teddy, author.
Title: Functions & variables / by Teddy Borth
Description: Minneapolis, Minnesota : Pop!, 2022 | Series: Coding basics | Includes online resources and index.
Identifiers: ISBN 9781532169632 (lib. bdg.) | ISBN 9781098240561 (ebook)
Subjects: LCSH: Functions--Juvenile literature. | Variables (Mathematics)--Juvenile literature. | Functional programming (Computer science)--Juvenile literature. | Mathematics--Juvenile literature. | Computer programming--Juvenile literature.
Classification: DDC 005.1--dc23

Hello! My name is

Cody Koala

Pop open this book and you'll find QR codes like this one, loaded with information, so you can learn even more!

Scan this code* and others like it while you read, or visit the website below to make this book pop.
popbooksonline.com/functions-variables

*Scanning QR codes requires a web-enabled smart device with a QR code reader app and a camera.

0011100100010011100100100110010100

Table of Contents

Chapter 1
Storing Data 4

Chapter 2
Code on a Mission 8

Chapter 3
What's the Function? . . . 10

Chapter 4
Functions Around Us. . . . 16

Making Connections 22
Glossary. 23
Index 24
Online Resources 24

Chapter 1

Storing Data

A program needs **data** to do a job. It can store data in variables. This data is called a value.

variable
variable
variable
COFFEE
TEA
SUGAR
Watch a video here!

variable
value

Variables are like jars. You put values into it. A cookie jar is a variable. The number of cookies is a value. You can change the value by eating cookies!

A program may need hundreds or even thousands of variables!

Chapter 2

Code on a Mission

Coders give names to variables. This makes the **code** easier to read. In basketball, each team's score is a variable. The number of points each team has is the value.

value

value

Learn more here!

Chapter 3

What's the Function?

Functions use variables. A function is a block of **code** that performs a job. The function of a seed is to grow. It needs **inputs** to work. The inputs are set to variables. The type of seed is an input.

Learn more here!

Adam waters his plant. Water is an input. Sunlight is too! The function takes the inputs. It runs the code and produces an **output**. The output is a healthy plant!

A function can get inputs from many places, even other functions!

input

output

The function can fail. The value for the water variable might be low. This is like getting an error in a code. The seed might stop growing.

Chapter 4

Functions Around Us

Chris is making pizza. The recipe is a function! The ingredients are variables. He uses cheese and tomatoes. The amounts used are values. The **output** is dinner!

Complete an activity here!

Amy plays a card game. The game uses a function to see who has the better card. The variables are the cards played. The values are the numbers on the cards. Amy's card has the higher value. She wins the round!

Values don't have to just be numbers. They can be types of objects or names, too. Eddy goes to the library. His function is to check out a book. The variable is the book. The value is the name of the book. He chose a coding book. Good choice, Eddy!

Making Connections

Text-to-Self

Have you ever made or ordered a pizza? What variables or ingredients did you pick? What variable tastes the best?

Text-to-Text

Have you read other books about functions or variables? What did you learn?

Text-to-World

Have you ever grown a plant? What variables do you think farmers need to grow something? Does the location of the farm change the variable?

Glossary

code – a list of instructions that tells a computer what to do.

coder – a person who builds programs or works with computer languages.

data – facts or items of information.

input – information that is put into a computer; what is put in or taken in.

output – the information stored in a computer when it is transmitted to a screen or printer; the amount produced in a given time period.

001110010001001110010010011001010000

Index

basketball, 8

code, 8, 12, 20

data, 4

functions, 10, 12, 14, 16, 19

gardening, 11, 12, 14

inputs, 10, 12–13

output, 12–13, 16

value, 4, 6–7, 8–9, 14, 16, 19–20

variable, 4–7, 8, 10, 14, 16, 19–20

Online Resources

popbooksonline.com

Thanks for reading this Cody Koala book!

Scan this code* and others like it in this book, or visit the website below to make this book pop!

*Scanning QR codes requires a web-enabled smart device with a QR code reader app and a camera.

110010001001110010010011001010000